I0481209

iOS 14 USERS

GUIDE:

The Complete and Simplified Manual on How To Use Apple iOS 14 With Easy Tips For Beginners And Advanced Users.

By

Ken Edward

Table of Contents

Table of Contents 2

Introducing The iOS 14 4

CHAPTER ONE 7

The iOS 14 7

CHAPTER TWO 14

Devices that support the iOS 14 14

CHAPTER THREE 20

The Compact design of iOS 14 20

CHAPTER FOUR 26

Messages... 26

CHAPTER FIVE 31

Maps.. 31

CHAPTER SIX.. 38

App Store in iOS 14 38

CHAPTER SEVEN 43

App Library 43

CHAPTER EIGHT... 48

Compact User Interface of iOS 14.................................... 48

CHAPTER NINE 56

Quick launcher in iOS 14 56

CHAPTER TEN 63

Translate App 63

CHAPTER ELEVEN 71

Safari 71

CHAPTER TWELVE 79

Car keys 79

CHAPTER THIRTEEN 86

Car play 86

CHAPTER FOURTEEN 94

AirPods 94

CHAPTER FIFTEEN 101

Camera 101

CONCLUSION 108

Introducing The iOS 14

All mobile devices make use of an operating system (OS) for their dynamic functionality. Many of these mobile devices use this operating system to aid the swift running of all other apps and software. Among many of these operating systems are Android, iPadOS and iOS which is majorly used for iPhone and iPod Touch devices. Apart from Android, iOS has been the most used OS for millions of people across the globe.

IOS is an operating system (OS) for mobile devices created by Apple inc. Since the introduction of this operating system into the market, Apple has made several adjustments and modifications to it to ensure maximum productivity especially in terms of usage. This modification has brought about a better feature which has been embedded in this iOS. This has made newer iOS to be better its performance and productivity.

The latest of the iOS operating system, the iOS 14, was recently released on the 16th of September, 2020. As such, there has been the introduction of much system adjustment, some of these adjustments include the ability to put a widget on the Homescreen and also change the default browser and other email apps. Aside from this, the iOS 14 has also been designed in such a way that Siri and phone calls now have a compact UI.

The recently released iOS 14 is seen as a successor to the former iOS, the iOS 13 which has also been used in many of Apple's iPhone and iPod Touch devices. Many of the other system features of this new iOS 14 are its App Clips which have a function resembling that of the App store but more dynamically, Carplay which allows users to use a built-in wallpaper, Car keys which enable the use of your device as a virtual car key by the use of NFC technology using compatible cars.

Many of its App features also include Camera, FaceTime, Home, Messages, Maps, Safari and Translate. With all this, you get to enjoy the comfort that comes along with taking a picture, texting a friend, seeking direction to a place, surfing the internet, using your favourite chatting app or translating from the English language to any other language of your choice.

Apart from these, you get to enjoy this new iOS 14 on any of Apple's iPhone devices that already supported the iOS 13 and particularly the 7th generation of Apple's iPod Touch device. This means that this iOS 14 will even offer a better function with any former Apple device without taking you through the stress of getting a new device.

You may have been wondering that you may not be able to adjust to the techniques of operating the new iOS that comes from Apple, we have compiled this book just for your consumption as we dive into the many unique features of iOS 14 and how to operate them. This will give you the details on how to run the latest iOS on the market.

CHAPTER ONE

The iOS 14

Figure 1 The iOS 14

IOS 14 is the latest operating system provided by Apple inc. for most of the users of its devices. Since its launch on the 16th of September, 2020, it has begun to become more popular among many users of Apple's devices like iPhone and iPod Touch.

Improved features of iOS 14

The launch of this OS has brought about many significant changes in the general function of many of the devices on which it is used. This iOS offers you many of the improved features that can be seen in the former iOS 13. Many of these improved features include

Unique experiences

- **Home Screen Widgets:** Widgets can now be added to your Home Screen and arranged in any way you like it
- **App Library:** The iOS 14 now arranges your app according to how often you use them. In this way, you can easily have access to your most-used apps
- **Compact calls:** All calls either from your FaceTime, iPhone or any other third-party app now appear on a designated part of your device without taking up the whole screen
- **Picture-in-picture:** This allows you to continue watching your videos or seeing your FaceTime call while you also can use any other app you want to use
1. **Better messaging**

- **Pinned conversation:** You can now pin about 9 conversational messages to easily find them when they need to use them arise
- **Group photos:** You can add a photo, Emoji or Memoji as a visual identity to your group conversation
- **Mentions:** To send a direct message to your friend, now can type the name of that person. You can as well receive notifications after you might have customized an active group for yourself
- **Inline replies:** You can now reply specifically to certain messages in a group conversation
- **New styles of Memojis and stickers:**iOS 14 now has several of these Memojis and stickers you can personally choose from

Maps

The new iOS 14 gives you a better experience, especially whenever you are seeking direction to or from a place. You can now get a better and easier route to such places just at the tap of your finger

- **Cycling directions:** Maps now let you get the directions of cycling by using the bike path, lanes and roads.
- **Guides:** Maps now offer you the guidance you need for most popular places around the globe. These guides can be saved and will usually get updated when more places are added

Translate

- **Conversation mode:** you can now converse with the new iOS 14 just by tapping on the single microphone
- **On-device mode:** You now get to use the on-device model to translate by making use of other necessary features of downloaded apps. This keeps your translations private, and you will not need to turn off the internet connection of your phone
- **Favourites:** You can now save translations in the Favorite tab to easily access them later when they are needed
- **Attention mode:** The iOS 14 now makes it easy for you to enlarge text for it to be easy to read

Siri

- **Compact design:** You now get quick information with the new compact design of Siri
- **Web answers:** Siri now searches the web to give specific answers to some question that you need
- **Send audio messages:** It is now easy to send audio messages especially when texting is time-consuming

Home Apps

- **Suggested automation:** The Home app now gives you some automation suggestions whenever you add a new HomeKit
- **Home status:** The most needed accessories in your Home app are now given preference so that you can easily access them

Safari

- **Web translation:** The Safari now can translate websites into 7 different languages
- **Password monitoring:** Safari give you a warning if you enter the wrong password

CarPlay and Car keys

- **Car keys:** You now can start your car with your device

- **Manage and share keys:** There is easy customization you can do to total or restricted access to any other person you want to share your Car Key with

- **CarPlay wallpapers and other new apps:** You can choose any wallpapers you need and add other new apps

Downloading and installation

Some people may encounter a little difficulty while trying to install the new iOS 14. To download and install the new iOS 14 is very easy. What you need to do is just to follow these simple steps

- Go to Settings on your device and press "General"
- Click on "Software Update"
- Press the "Download and Install" option

Usually, it takes more than 15 minutes for the updates to get installed. So, during this time you will be unable to use your device. It is, therefore, necessary that you perform the downloading and installation when you have enough time.

CHAPTER TWO

Devices that support the iOS 14

Figure 2 Devices that support the iOS 14

For many users of the Apple's iPhone and iPod Touch product who have been wondering if their current device may be able to work with the just-released iOS 14. The iOS 14 has been made in such a way that it will be able to work with some older versions of Apple's iPhone devices. Here is a list of devices that can run this OS

- iPod Touch (seventh generation)

- iPhone SE (second generation)
- iPhone XS, XS Max
- iPhone XR
- iPhone X
- iPhone 8, 8 Plus
- iPhone 7, 7 Plus
- iPhone 11
- iPhone 6S, 6S Plus
- iPhone SE (first generation)
- iPhone 11 Pro, Pro Max

Modifications in iOS 14 Home Screen

The main change that has taken place in the new iOS 14 is that users now can add their specific apps to the Home Screen. Users get to add or delete such app widgets at their convenience. This has not been so in the former versions of iOS. The former versions were made in such a way that the icons that appear on the Home Screen have been pre-arranged in an orderly way and these icons correspond to the specific app.

With this new modification, users can now hide some apps that are not often used and maximize those that are often used. This makes these apps to be organized in the new App Library. You can as well customize your Home Screen using the on-device app like the Shortcuts app or a third-party app like Widgetsmith.

Customizing your Home Screen using Shortcuts

- Go to the Shortcuts app on your device
- Press on the "plus" button on the top-right corner of your device
- Choose "Add Action"
- Go to the search bar and type "Open App", then, choose the Open App
- Press "Choose" to choose the app you wish to customize
- Press on the 3 dots on the top-right corner and give a name to what you have created
- Press "Add to Home Screen"
- Name the shortcut to what you like when your device asks for the Home Screen Name and Icon

- Go to the Safari app to see the new icon image. Save this image to your photos
- Go back again to the Shortcuts app and press the current icon
- Select "Choose Photo" and press on the image you saved
- Press "Choose"
- Press "Add"

Customizing your Home Screen using Widgetsmith

- Download the Widgetsmith app
- Choose the size of the widget you wish to customize. Either small, medium or large
- Press on the widget. You also get to change the font and colours, then, press "Save" when you are done
- Go to Home Screen and long-press on an app
- Press "Edit Home Screen"
- Press the "plus" button on the upper left corner of your device, look for Widgetsmith and press on it.

- Choose the widget size you wish to add and press "Add Widget"
- To alter this widget, just long-press on it and choose "Edit widget"

Modifications in iOS 14 Widgets

One of the main features of the Widget in the new iOS is that it can be put on the Home Screen and be among other app icons. Aside from this, a scrollable widget UI has replaced the Today View which is at the left of the first page of your device.

These widgets can now be resized into either two by 2 by 2 icons, 4 by 4 icons or a horizontal 2 by 4 icon. If these widgets now have the same size, they can be arranged on top of each other and swiped easily.

Apart from this, you can now use the Smart Stack platform to arrange your widget in such a way that the recently used widgets will be given preference over others so that you can easily access them. If you have been wondering how to about this, then follow

Adding and removing

- Press on your device for a few minutes till it enters the "jiggle" mode or just press the "Edit Home Screen"

- Press the "plus" sign on the up-left corner of your device. This will make the Widgets menu appear

- You can now search through the lists of widgets you want to add.

- If you have chosen the widget design you want, then press "Add Widget" which is at the bottom of the screen

- The widget then appears on your Home Screen. You can further edit its position here.

- To remove a widget, just press on the "minus" button when in the jiggle mode or tap on the widget for a few seconds, then choose the "Remove Widget" option from the menu.

CHAPTER THREE

The Compact design of iOS 14

Figure 3 The Compact design of iOS 14

The iOS 14 has now been designed in such a way that multiple actions can be performed without interfering with any other task that is been initially done. You may be in the middle of a task and your phone suddenly rings, in the former design, you notice that you will not be able to continue your task except you pick up your call. Now, you can perform but he functions simultaneously.

This occurs because a notification banner is now shown for incoming phone calls rather than being displayed on the full screen of your device which is seen in the former versions of Apple's iOS, like in iOS 13.

This same features so happen whenever you want to receive a FaceTime call. But when your device is locked and not in use. To decline any phone call, just press on the reject button which is on the coming call banner. If you, however, press on the accept call button within the incoming call banner, the call will be activated. If you, however, tap outside it, you will have the incoming call as a full screen on your device.

You can also decide to swipe the incoming call's banner away and continue to use your device. If you do this, the incoming call will ring on the background with its small icon which indicates that you have a call.

Opening the full screen for any incoming call, therefore, mean that you will have no option but to accept or reject the call for you to continue to use your device. But if the incoming call's banner has been swiped away, you can always press on its icon to go back to the calls display. Other features of this compact design are

Picture in Picture

This is another compact design for Apple's iOS 14. With this new feature, you can now watch videos and other TV shows from your device's app or on the web while doing another task on your device. This is because Apple has made a provision for you to do any other task in a window view while watching your favourite movie or TV shows.

Another interesting thing is the FaceTime picture-in-picture mode. This allows you to do another task while having a FaceTime conversation with a friend or loved one. Unlike the iOS 13 that will pause your FaceTime conversation when you exit, the iOS 14 will only collapse it into a small window and allow you to continue your discussion while doing other things.

Siri

For people who often use Siri for quick answers to some questions, when you now activate Siri on iOS 14, it now comes up as a small icon at the bottom of your device. You can be sure that Siri is listening to your questions because its icon moves as you ask your question.

As you speak or request several answers from Siri, its icon moves to indicate that the voice assistant is listening and picking up whatever you are saying. A banner format version of your queries is also displayed in a compact view at the upper part of your device with all the necessary information. Your information is also given in a small banner form at the upper part of the screen of your device. The longer the information sourced by Siri, the longer Siri's banner and the more space it takes on your device.

Translation support and smartness of Siri

The iOS 14 Siri is now dubbed the smartest. This is because Siri now has more facts and information before. Siri can answer much more direct question which is satisfactory. Siri has been made about 20X better than previous versions and can provide answer approximately 25 billion questions monthly.

Apart from this, Siri can now translate other languages into the English language. The major languages of translation supported by Siri are

- Portuguese (Brazil)
- German (Germany)
- Chinese (Mandarin, Simplified)
- Japanese
- Russian
- French (France)
- Italian (Italy)
- Arabic.
- Spanish (Spain)

- Korean

CHAPTER FOUR

Messages

Figure 4 Messages

Apart from the Calls and FaceTime conversations of the iOS 14, this OS also supports Message chats or conversations, thereby allowing you to customize your messaging app just the way you like. Some of the things you can do on the messaging app of the iOS 14 include

Conversations and pinning

You can pin your most-loved conversation as the first in your conversation lists. This enables you to access them easily when they are needed. Besides, a group conversation can also be pinned and the three most recent members who send messages to the group will be circled around the pin. Pinning conversations help you to easily get access to your most important messages rather than looking for them when the need arises.

To pin any of your conversations, just swipe on the conversation towards the right or tap on the conversation for a few seconds, then, select the "Pin" option. To unpin any of the conversations, just tap on the pinned conversation for a few seconds.

Customizing Memojis

With the new iOS 14, customizing your Memoji is now an exciting experience. You now get to enjoy more improved features of your Memoji and you can change it to the exact Memoji you want. This improved Memoji features

- Eleven new hairstyles which help you take the perfect hairstyle you want in your Memoji
- Nineteen new headwear styles which let you showcase your hobbies, profession or religion
- Three fresh stickers for Memoji which you can use to send a blush, hug or fist bump to a friend

- Six new provisions for more age options that allow you to choose the exact age-look you want in your Memoji.

- Better Memoji expressionwhich let you choose the best Memoji similar in personality and opinions with you. This has been specifically improved by Apple so that your Memoji have better facial appearance and muscle which contributes to more expressiveness

- Face coverings which enable you to choose the look you want. You get to select your favouritecolour that best suit you.

All these features will make your Memoji take the place of your physical appearance and your person can easily be known to others. With these Memoji customizations, you get to change your Memoji to the perfect you.

Groups

You will also be able to send a direct message to any person within the group by typing the name of such a person. The most active of all the groups may be customized in such a way that makes you get notified whenever you are mentioned. To create a mention in your group messages,

- Go to the "Messages" app from your Home Screen
- Select the group chats
- Input your message
- Type "@ the person's name" to create a mention (I.e "@steve" if your message is to be directed to steve)

There are other times when your name will be mentioned. In this case, you will always be alerted even when you have turned off necessary notifications. You can, however, change this from the Settings of your device. To do this

- Go to the Settings app on your device
- Select "Messages"
- The "Notify Me" option under Mentions should be toggled off

Aside mentions, the iOS 14 has been made in such a way that you can now keep track of many replies and as well reply messages sent earlier by another person. .This ability of the iOS 14 has been dubbed "in-line" replies. To give an in-line reply,

- Press on any message you want to reply for a few seconds
- Select reply

After this has been done, your message will show up as the most recent message but the original message you replied to will also be seen by users who are also using the latestiOS 14. This simply helps you to monitor your important group messages that need your attention.

You can as well set any image you like for the group conversation, which may be either a photo, Emoji or Memoji, which will be shared with all members of the group.Another remarkable thing done by Apple was improving the interface for the group chat. The collections of people profile pictures can be seen at the top of the chat. This will show in such a way that the most active participant's picture is shown the largest.

CHAPTER FIVE

Maps

Figure 5 Maps

The improved iOS 14 maps let you search for things like addresses, landmarks, places etc. Unlike the iOS 13 in which there are limited data about some places you can search for, the iOS 14 has been updated with a lot of spectacular information about many places, especially across the United States. Some of the most important features embedded in the Maps of iOS 14 include searches for cycling routes, the map guide, Electric Vehicle routing and CarPlay.

How to find amazing places using iOS 14 maps

Finding a place of your choice using the iOS 14 map has been made a lot better and easier. This is because, with the improved features of iOS 14, you get to know as many locations you wish to visit. To search for any place of your choice in the Map, you may decide to ask Siri or simply do that by taking the following steps

- Go to the "Map" app on your device
- Type the name of the place you are looking for on the search box
- Press on location in to see it on the map

    If you are intending to go towards the same place that you searched, then you can select "Directions" towards your path. This will immediately update you on the particular route to take towards the place. Aside from this, you can as well look for locations that are near you using the Apple Maps, to do this

- Select the location icon at the upper right corner. This let the map be on your current location
- Tap on the search box and select any category you want
- Press on location option from the category you selected
- Select "Direction" to get the necessary information on how to get to the place

Your current location should be turned-on on your device by going to "settings", press on "privacy", the "location services". Press on the map, then let your precise location be on. With this, you can always get the direction towards any of the places you searched for.

Cycling in Maps

Apple Maps also allow its users to use many cycling directions. You now get to know which way to go especially when cycling to avoid the risk of passing through a busy road or hilly place. You can use cycling as your preferred travelling path;

this can always be done on the "settings" of your device.

Some features for cyclist

- **Routing:** Maps can now let cyclist be routed along bike paths, bike lanes and roads that will not offer difficulty in travelling for most cyclist. You will usually get to know about the need for you to get off your bike when you are approaching any difficult path especially elevations that can only be walked. This will majorly ensure that you have an easy ride.

- **Preview:** Maps offer a preview to what will be on the road while cycling to avoid unnecessary cycling disturbance. You get to see several warning labels on your Maps. You will know if there is a steep road or stairs ahead of your journey

- **Voice guidance:** There is now a cycling voice guidance which will assist while cycling. This makes it easy for you to navigate your way. You will no longer need to have your driving instructions with you when cycling again.

Using Maps for cycling direction

- Go to Map and use the search bar to see the direction
- Press on the new bicycle image to change your routing. You will now see the distance and approximate time for travelling
- Swipe up on direction card to toggle to avoid hills, steeps and busy roads
- Press the time estimate/elevation graph to see the details of your route before starting the journey
- Press 'Go" whenever you are ready for the journey

Guides in Maps

You may be looking for a particular place to have your lunch or a favourite relaxation centre. With Apple's guide, you can now seamlessly locate the place. Apple, in conjunction with many travel and tourism agency, has created this feature. For instance, if you are looking for the best pizza shop in San Francisco, you will to browse it out and get easy access to their pages on Maps. This guide can be saved for viewing later. To use the guide,

Go to Maps

- Search for your place of choice
- Scroll down the option that appears and press on any two or three listed Guides
- You can also press "see more" to see other available Guides
- You can now read every information you need through the Guide

Apple Maps will highlight all places around your search word and give you details concerning what they do. You can as well use the direction to get to the place. To close the Guide, simply press the X sign (close box) that is at the upper right corner of the Guide.

Electric vehicle routing

The advent of electric cars has made Apple improve its Maps app by using electric car routing. This allows the Maps to know the model of your car and then the destination you are going to. The Maps let you know where to stop and top-up your electric vehicle. You will also get to know the timeline for which your car will be fully charged.

This is particularly good news for most users of BMW and Ford because these two carmakers have already partnered with Apple on this. This will now make many of your journey easy as you will always get updated about both the conditions of the road as well as your car.

CarPlay

The Apple CarPlayis a feature that can be configured with the Map in such a way that you can now be at alert to speed traffic, cameras or the red light. This also enables you to know the congested area along your journey to easily take another route. You can now drive with confidence and less difficulty to your destination

CHAPTER SIX

App Store in iOS 14

Figure 6 App Store in iOS 14

The Apple App Store will allow you to download and use any of the apps developed by Apple. You get apps ranging from as social media apps (Messenger, Instagram, YouTube, Uber, Snapchat, Netflix etc) to gaming apps (Subway surfers, Rolling sky, Fortnite, Mario Kart Tour etc). This is the hub to download any app you find useful. Some of the new features of App Stores are

- **SKOverlay:** This makes you download without having to leave the current app you are working with. You just need to press on the overlay if you want to view the app you are currently downloading

- **Management of your sandbox account:** You now can manage your sandbox account which gives you the permission for subscription, test upgrades and downgrades, introductory offers and cancellations. This you can always by pressing setting, then sandbox account
- **Access to UserInfo parameter:** You can now pass your UserInfo to Account Authentication Modification Extensions when doing an in-app upgrade. This allows for easy usage of the UserInfo parameter.

App Clips in iOS 14

Apple introduced App Clip for most of its users who find it stressful having to download various apps from the App Store. App Clips enable you to use a part of any app without the need to download the app.

To use the App Clips, you will need to search for a business or website that is compatible with it and then press on the App Clips button. This allows you to access the web-like version of the app and see what such app offer before you download it.

Notes in iOS 14

Apple Notes is not new in the world of iOS. However, there has been a little bit of improvement for the Notes app in iOS 14. Notes is a very useful tool especially when you need to quickly save an important session or document. Apple note particularly makes you to easily save text or other media notes from your iOS. Some of the features of the iOS 14 Notes are

- **New actions menu:** This menu appears when you are in a folder or note. You press on the menu (three dots) at the upper right corner to access the actions menu. With this, you can view attachments, switch to gallery view, sort notes or delete and move multiple notes. You can as well scan, pin, lock or delete any chosen note.

- **Pin Notes:** You can pin note for you to access them later when they are needed. To do this, you just have to press on the note for few seconds, then press on "pin note". You can as well do this by opening the action menu and selecting "pin note".

- **Quick styles:** You easily get to switch to the style you want when typing any important document on the Notes app. You can press "Aa" for a few seconds and quickly change the title, heading and body of your notes to what you want. You can as well change your text to bullets or numbers. Hold down your finger, and then slide up to the quick style option you want. After this, release your fingers. You will immediately get you desired note style

Notifications in iOS 14

Notification is a vital part of any device. The Apple's iOS 14 Notification enables you to keep a record of what is latest happening on your device such as a missed call, message, upcoming event etc so that you can always respond quickly to such notifications. Although when your device is not in use, notifications are seen on the lock screen, the Notification centre also allows you to see all notifications.

You can also change your notification settings to enable only important notifications. This can be done by going to the setting app. With this, you can turn on or turn off notifications, allow notification to be visible either in lock screen or notification centre, remove notification etc. To see your notifications,

- Swipe up from the middle screen of your device when it is locked or
- Swipe down from the upper centre of your device when it is unlocked. This allows you to see any other notifications you may be having

CHAPTER SEVEN

App Library

Figure 7 App Library

This is one of the latest additions to Apple's iOS 14. The former iOS introduced have no Library for all the apps. Hence, many installed apps usually have their icons on the Home screen, which gives the Home screen a congested look.

Without the availability of the App Library, you will always have to scroll through your Home screen before you can find any of the apps you want to use. Apple has changed this by making a unique separate place where your entire apps can be seen in an organized fashion. This is where you get to even customize your app categories.

The main function of the App Library is to put all of your apps into one view portion and allowing you to see any of such apps at a glance. By doing this, your apps become very easy to access and use at any point in time. The App Library can be found at the end of your Home screen (i.eat the right of the last page that holds your apps). Some other few functions of the App Library are.

Automatic categorization by the App Library

This feature of the App Library enables your apps to be arranged automatically into different categories which relate to the function of each app. Aside from this, they also arrange such apps according to how often they are being used. Your apps may be arranged into categories like Productivity, Entertainment, Creativity, Social or Recently Added.If you want to use the App Library,

• Go to the App Library page by swiping left to the last page of widgets or apps on your device

- Press on the App Library (there are different categories here with three large icons of apps and a collage at the down-right)
- If you press on the app icon (i.ethe large icons), you will instantly have access to the specific app you want from a list of other small icons
- Press the small icons to open the type of category you want
- Look for the app you wish to open and press on it

Although, all the apps on your device will usually be found in the App Library. If you do not want some of these apps to appear on your Home screen, you can just put them in the App Library alone (this means that you will not be able to see them on any part of your Home screen) and if you want to access them, you can just go to the App Library. To do this,

- Long-press the app for few seconds to see the menu option
- Press on "Remove app" from the menu option
- Press "Add to Library"

Another thing you can do with the App Library is to allow apps that are downloaded to show only in the App Library. By doing this, the downloaded apps will not appear on the Home screen of your device. To do this,

- Go to the Settings of your device
- Press on the Home screen
- Press the "App Library only", this will let newly downloaded apps to be in the App Library alone
- Press "Show in App Library". You can switch on or switch off notifications, depending on whether you want notification badges to appear in the App Library

Suggestions by the App Library

This is another feature of the App Library which makes it to bring out some suggested apps for you whenever you are using your device. These suggestions are usually based on the location of the app, the functions of the app or how long you have been using the app.

Most time, the App Library will give you the suggestions to either the recently used app or your most used app. In this way, you get to quickly open any of the apps that you are constantly using without any need to go to your Home screen.

Searches by the App Library

A search bar is found at the topmost part of the App Library which makes it better for you to easily find any app you want to search. When you press on this search bar, it automatically displays some of your apps alphabetically.

You can scroll through the displayed apps and choose from the list of the suggested app any of them you wish to use.Otherwise, you can as well just type the name of the app you are looking for in the search bar, the app will come up from the suggestions list along with other related apps whose search letters are similar and you can then press on the specific app you want and open it.

CHAPTER EIGHT

Compact User Interface of iOS 14

Figure 8 Compact User Interface of iOS 14

The former iOS has featured a full-screen for all incoming calls. This means that you will always answer you call even when you are doing any other task with your phone or you can otherwise wait till the call stops for you to continue to use your device

The new compact user's interface is like a banner which is at the top of the screen of your device. It let you do things like silencing calls with a swipe, answering a call while keep doing what you were doing. You can also expand this compact UI to the full-screen size for the several other options in the Phone app or third-party

Phone

Compact call interface is set as default on devices with iOS 14. You can always answer or decline a call from the compact UI which is at the top of your screen. The green button will answer while the red button will decline it.

After you might have answered your calls, you can continue to use your device as you have been doing without leaving the Phone app (Press on the green phone icon at the top left corner to go to your Phone app or any other third-party app)

You can swipe up to silence the call without declining it

- The call icon is at the upper left corner of the screen of your device in case you change your mind and decide to pick your call

- If you press or swipe down the compact call interface, it will be back to fullscreen (to get audio output options, dial pad, etc.)
- If your device is locked (screen off), the full-screen alert will be seen

After answering your call using the compact UI, it will slide away by itself after some seconds. You can as well swipe up if you want it to go away immediately while you continue your call). Press on the call image which is at the upper-left of your device to go to the Phone app.

FaceTime call

If you have been particularly keen on minimizing pop-ups of your device call, then, the FaceTime call of iOS 14 offers you that easy compact UI design that makes you focus on your calls and at the same time disallowing the takeover of your device's screen by such call.

When you are receiving any incoming FaceTime call on your device, it now shows just like a banner which you can decide to pick, ignore or just swipe it away. This usually occurs when your device is being used. If your device is not in use and is locked, then the FaceTime camera will be activated and the call will take up the entire screen of your device.

Whenever you decide to swipe away the call banner, the call will continue to ring in the background. This will however not disturb you especially if you are in the middle of some urgent task. You can continue your task in such a case.

The FaceTime call icon ringing in the background will be seen at the top left corner of the display in this case and can be tapped upon when you decide again to accept such call. If not it will continue to ring till it stops. If you press on the FaceTime call at any moment in time, it will expand fully on the screen of your device. This will either force you to accept the call, reject it or just ignore it till it stops.

Compact Siri

This is another compact UI design by Apple which will make the activation of Siri not to overtake the whole of your screen. When Siri is activated in this case, a small icon (which indicates that Siri has been allowed) comes up at the bottom of your device. When you speak into the Siri voice command, this icons move indicating that Siri can hear you and it is surely assisting you for any question you may be asking.

Many of your requests are usually shown are a compact view and the provided information which Siri has sorted out will be seen in the banner pop-up which is at the upper part of the display.Although you are always sure to get your information from Siri, however, longer information may take up more banner-style spaces on your device especially those involving web searches.

Siri will not also display the texts of the questions you are asking because of its compact view but you may be able to turn this on by enabling the "Always Show Speech" feature. When this is done, whatever question you may be asking Siri will only show up as a small pop-up on the Siri icon. This especially good if you want to make sure that what you are asking is being interpreted correctly by the voice assistant. To put on the features of this "Always Show Speech"

- Go to Settings
- Go to Siri and search
- Press on Siri Feedback
- Toggle over the option for Speech for the "Always Show Speech" feature

Other features of the compact UI design for Siri also allow Siri to support sending audio messages, sharing Apple Maps ETA and getting cycling directions. However, if the Siri interface is open and you are attempting to press on any other place on your device, the Siri interface will close. For you to start another Siri request, you will then have to press on the Siri icon.

Third-party VOIP calls

Third-party VOIP can now integrate their apps into a new API created by Apple for this purpose. This new API is for collapsed call interface and it helps most developers of a third party VOIP apps (like WhatsApp, Skype, Facebook Messenger etc) to merge the features of their apps into the new API. It is, therefore, necessary for most app developers to allow the API into their app. If not, their apps will not be shown in the call interface of this API.

Usually, you will need to enable any such third party VOIP app for them to function with your device. This is done by enabling the app in the iOS Settings to work with calls especially if it a Siri command voice prompts. This allows you to use the third party app right away from the voice prompt/call of Siri without launching the specific app. This is particularly good if you are performing an important duty at that specific time.

You can start many calls on such third-party apps by using the call prompt of Siri. For instance, if you are using WhatsApp or Skype, you can simply say to Siri that "Call ....(your friend's name will be here) using WhatsApp" or "Call ....(your friend's name) using Skype). Many incoming calls are also supported by this compact feature of third party VOIP of iOS 14. This has been made possible by its developer API.

CHAPTER NINE

Quick launcher in iOS 14

Figure 9Quick launcher in iOS 14

The quick launcher feature of iOS 14 allows you to easily launch an app or go to websites when you input a few characters.You just have to press the "Go" and any of such inputted word which represents the app you want to launch will automatically launch itself. This offers easy accessibility to most apps on your device as you will be able to quickly open or launch them.

One of the other interesting features of this quick launcher is that you can just press the back of your device to launch certain apps. This provides you with comfort especially when your device has been locked, you can just simply tap at the back of your device and go straight to the specific app you wish to open.

This is done by double-clicking or triples clicking at the back of your device and such app will respond according to the preference of your device's settings. This feature of the quick launch is popularly called "Back Tap". This feature can be set by going to

- Settings
- Accessibility
- Touch
- Back Tap

The in-app search of iOS 14

Many apps such as Messages, Files and Mails allow you to search out things from other lists of suggestions.This is a good experience if you need certain information that is contained within any of the apps on your device. You can simply by-pass any other mean of searching and just open up the app you wish to open. This in-app search does this by allowing links to automatically open your apps when such links are pressed.

The in-app search performs just like any other searching means. It permits users to input their requests and then they can easily see the search result for themselves. For instance, you may type a friend's name using the in-app. This may take you to the Facebook profile of that friend and clicking on the pop-up link that accompanies that in-app search will automatically open the Facebook app on your device.

As-you-type search suggestions in iOS 14

This feature of searching in iOS 14 allows you to type whatever you may be looking for into the searching option. Several other suggestions will then come upwhich will allow you to choose what you are looking for from the lists of the drop-down suggestions. To search, simply

- Swipe down the Home screen of your device from the middle

- Press on the search field
- Input what you are looking for on the search field (as you do this some suggestions comes up)
- Press "Show more" if you want to see more of the search results or you can just press "Search in app" to directly search from any app
- Press on any of the search results that you are looking for to open it

  Another feature of the as-you-type search suggestions is that some of these suggested words canbe auto-corrected in case there is a miss-spelt word. This auto-correction also offers an easy typing experience especially if you want to avoid the rigours of typing and then clearing word and retyping them again. If you are keen on using the as-you-type search suggestions for auto-correction, you can simply just

- Go to settings
- Press General, then, Keyboard
- Turn on "Auto-correction"

The search can also be set in such a way that some apps which you will want to appear when you are searching will be limited in number. This enables concise search and by doing this you will easily get faster access to what you are looking for. To change this,

- Go to Settings
- Press on Siri and search
- Choose Suggest app, show app in search, suggest shortcuts for apps or show content in search
- Turn this off by going to Settings > Siri and search > turn off such settings

Top hit results

The "Top hit results" of the iOS 14 features the most important results from a range of many other results. These results are usually seen in the upper part of your device. Most times, results such as Maps, apps, contacts, websites etc are shown. This allows them to be easy to find especially when you need them urgently.

In case of web viewing especially when you are using web browsers such as Safari, the top hit results will usually show as the first set of results for your search. This will particularly be pre-loaded already on such web browser so that as soon as you press on it, the links of the results open faster. This saves your time as such a website will be accessed within the shortest time possible because they have been preloaded. To turn off these top hit results,

- Go to settings
- Press on Safari
- Turn off the "preload top hit"

Web search

To now search using the web is easier with the redesigned feature of iOS 14. You can simply type and see related websites and other suggestions of web-related searches at the upper bar of your device. This will make launching Safari to be easier and you can just go to the full web version.

Previous searches that you have made on the web may also appear on the search bar.

The Home screen search tool has even been made better for users of the Safari browser. This is because you can use your Safari to quickly search for apps, web information, text messages, settings and many other things. You can now switch your default browser to any other browser but you are always sure to get many lists of possible suggestions as you browse through the web.

Search engines like Safari will search through any content in the entire web and keep many of such data on the different pages of the web that has been catalogued. This easily makes much information to be available for you just at the press on your fingertips.

CHAPTER TEN

Translate App

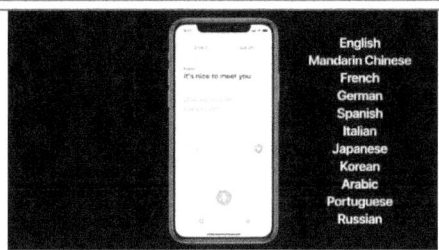

Figure 10 Translate App

Siri can always translate languages but you'd have to ask Siri. The iOS14 comes better with a new Translate App feature where you can do all your translations. The translate comes as a standalone app so that translating will now go smoothly in a conversation and you don't have to ask Siri. The Translate app uses a neural engine to translate words or phrases and it offers a total of 11 languages that you translate to.

The Translate app has some useful features such as the conversation mode, offline mode, attention mode and you can also have a favorite translation. You can translate using text or speech in the desired languages you choose on the app. The features of the app make it even more useful as you can have a realtime conversation with a person using the conservation mode of the app.

The default language to translate from on the app is the default language of your phone and the default language to translate to is Spanish. You can change this in the app if you want to translate from another to the preferred language you want to translate to.

Using the app

The Translate app can be used in portrait or landscape. Available options are slightly different from each other depending on which mode you are in.

In portrait mode, the display of the phone is split into sections. The upper section shows the language you are translating from, the language you wanted to translate to, the language you wanted to translate to, the phrase you translated, and its translation. The lower section shows where you can enter more text or use the microphone to record a phrase you can translate from.

In landscape mode, this focuses on the microphone as means of inputting text for translation and this is where you can access the attention mode feature of the Translate app which you might prefer when you are in conversation with someone.

Voice translation

The voice translation feature lets you speak to the app so that it gives you the translation. You can speak in phrases or long sentences for the app to translate. The Translate app supports 11 languages so you can speak in any of the supported languages for the app to translate it to you in the other language.

- Launch the app
- Change the language you want to speak in and the one you wanted to translate to in the upper boxes
- Tap the mic icon to speak to the device

- The Translate app gives you back the translation in the targeted language.
- To translate another phrase or word, tap the mic icon again to translate the it.

Text translation

The text translation lets you enter your text, phrases, or words you want to translate into the app. This works in the portrait mode and it is useful when you need to paste a text or something from a webpage or document for translation.

- Launch the app
- Change the language you want to translate from in the upper left corner this is set to the default language of your phone.
- Set the language you want to translate to at the upper right and corner.
- Tap the "Enter text" in the app and it shows where you will enter your text or paste the text you want to translate.
- After inputting your text, tap the go button on your keyboard.

- The Translate app will give you the equivalent translation of the words you entered in the language that you choose, this will show below the words you entered into the app.
- You can listen to the translation by tapping the play button at the bottom right corner of the app to get the right pronunciation in the language.
- To enter another phrase or words just tap the "Enter text" in the app to show the interface where you will enter your text to give you the translation.

Conversation mode

This is a feature that lets you chat back and forth someone that speaks in another language. To get to the conversation mode you have to turn your device into the landscape mode. The device listens for both languages and translates between them which shows on the screen of the device. To use the conservation you should be tapping the mic icon in between conversations with the other person. When each person talks the iPhone translates it.

The conversation mode can also be used with the automatic speech detection feature. To use this make sure you toggle on the auto-detection by going to the language at the top and scroll down to toggle on the Automatic detection. If Automatic detection isn't working well you can toggle it off and use the mic icon in-between speeches instead.

Favorites

In the Translate app, transactions can be saved as favorites if you know you use them often. Through the favorites tab, you can save any recent translation as a favorite for easy accessibility. The favorite tab will also show your recent translations.

Attention mode

When in conversation mode on your device, tap the expand icon (two outward-facing arrows) on the screen to enter into the Attention mode. This let's the translated to show in a bigger font that takes up the entire screen. This is particularly useful when you need to get the message across to a person some paces away when you can't speak the language.

The Attention mode is better with short phrases rather than long sentences because of the zoom. Tapping the play button will speak the translation aloud to get the pronunciation and tapping the conversation bubble icon leaves the Attention mode back to the Conversation mode.

Supported languages

The Translate app supports 11 languages that you can translate from and to. These languages are also available offline download so you are able to do an on-device translation. To download the language packs, tap on one of the language boxes and tap the download button next to the language you want to download. The supported languages are listed below.

- Arabic
- Chinese
- English (either US or UK)
- French

- German
- Italian
- Japanese
- Korean
- Portuguese (Brazil)
- Russian
- Spanish

CHAPTER ELEVEN

Safari

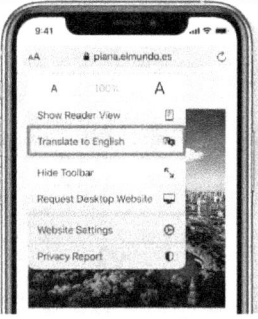

Figure 11 Safari

In iOS 14 Apple does not leave the Safari browser lagging behind. The browser is taken a notch higher than before with upgrades and a host of added functionalities and features. It has some truly exciting features that enhance the performance of the browser though there are no changes to its aesthetics the browser has some privacy oriented additions, performance enhancement upgrades, and a host of other added functionalities. These upgrades make the browser better than ever before.

Improved performance

The Safari is given a boost ahead of its rivals in this new version of iOS. Though the aesthetics are basically the same, it now has a power-packed powerful under the hood boost to it's JavaScript engine. This boost to its engine enhances the performance and clocks twice the performance of its main rival browser Google Chrome on Android. Sites are now expected to load faster and also some site elements are expected to respond faster.

Hopefully, third-party web browsers (that are based on safari rendering) will take advantage of this increase in performance as well.

Web page translation

Safari web browser now translates web pages in 7 languages. These features eliminate the need for a third-party app to translate or copying words from the web page to the Translate app. The translations feature is very useful as you can now access websites that are not primarily written in your default language. You can now translate webpages by the tap of an icon.

The built-in translation option can be used to translate websites into 7 languages, it can translate the webpage to English, French, German, Chinese, Spanish, Russian, or Brazillian Portuguese. When you visited a website that the safari can translate an icon appears (aA) at the top of the browser.

The translation feature is set to your default language but you can select other language options to add them to your preferred language list so you can translate to it when you visited a webpage. The language options can be selected from the settings app on your device.

To add other languages to your preferred language list

Websites can be translated into additional language If you have selected them before to your preferred language list

To update your preferred language list

- Launch the Settings app on your device
- Tap General -> Language and region
- Tap other languages
- Select the additional language from the list.
- Tap keep (Your preferred language) to keep your default language that the device is using and confirm the new selection as an additional webpage translation language option.

Following these steps will allow you to choose additional languages to translate webpages too. If you visit a website that can be translated to that language it will be available as an additional translation option.

Using the translation feature of the Safari web browser

Launch the Safari app on your device and visit a webpage you'd like to translate.
If the webpage can be translated an icon will appear at the far left of the address bar.

Tap the icon (**aA**) then select Translate to (Language) in the dropdown menu. If the option isn't showing it's either the webpage isn't compatible with the safari translation feature or the language isn't supported.

Tap **Enable Translation** if it shows in a prompt.

To view the original language of the webpage back. Go the icon and tap on View original. There is also an option to **Report Translation Issues** if you see one.

Password Monitoring

Safari iOS 14 is can watch all saved passwords closely, and alert you if they are being involved in a data breach. The cryptographic techniques that Safari uses, helps check your password derivations against a list of breached passwords which was promised by Apple to be a secure and private way.

The feature is added because data breaches can expose your login credentials and pose significant threats to your security and privacy. If there is a data is a breach you wouldn't know your username and password has been compromised until something bad happened.

If there is a breach, Safari alerts you and gives you a prompt to sign in with your Apple ID on sites that support the feature or it generates a new and secured password automatically.

Potential problems with passwords can be seen under Security recommendations in the Setting app in your device.

Website Privacy report

Apple adds a privacy report feature in iOS 14 that enhances the Apples Intelligence Tracking Prevention Functionality. Work is being done to prevent cross-site tracking, which makes website monitor the internet usage on your device as you browse various websites for analytics, ad targeting, and so on.

 Intelligent Tracking Prevention is the description from Apple for various tools that keeps tab on cross-site trackers and block them in Safari. The privacy report list sites that are using trackers, the amount of trackers installed on those sites, and the prevailing tracker you come across when browsing the web.

Browsing through the web, you would find trackers installed on a website that uses ads for monetization, as well as an ad network. Also, sites that analytics services to monitor your web behavior and collect data for content and site improvement.

The privacy report in safari shows ranges information and list them out for you. Safari shows the trackers in their numbers for each site you visits, the number of teachers Safari has prevented, numbers of sires visited that have trackers installed, and a list of frequently visited sites that have trackers installed like Google's doubleclick.com.

You can access the privacy report section in safari via the website view menu by tapping the icon with two A's next to the address bar and choosing a privacy report. This will bring out the list all cross-site trackers that are monitoring your activity over the web.

CHAPTER TWELVE

Car keys

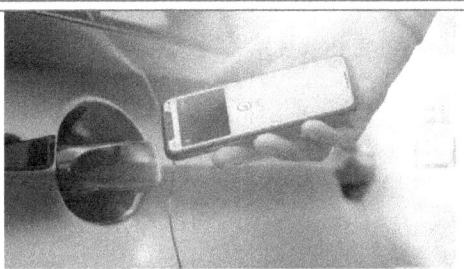

Figure 12 Car keys

A new feature comes with the iOS 14, in which your iPhone will turn to your car key. This feature uses NFC (Near Field Communication) for communication between your device and the car. The feature lets you start your car, unlock your car and it has a whole host of other functions which include sharing your car key, remove car key access when you share it, and set the car key type you are sharing to modify its functions.

The iPhone is paired up with your car assuming your car support this feature. The digital key communicates with your car via the NFC and it is stored in the same secure place you store your credit cards. The digital key is stored in your wallet means that it is protected by Face ID or Touch ID.

The BMW series 5 is the first to support the Apple car key feature. Not all iPhones that uses iOS 14 can use the car key feature. The compatible devices are iPhone XR, iPhone XS, iPhone XS Max, iPhone 11, iPhone 11 pro, iPhone pro max, iPhone SE (2nd gen), Apple Watch series 5.

Apple ensures that the process is safe and secure based on a unique token that is shared between the car and the iPhone. The car key can also be added to the Apple watch which makes it even easier to use. The digital key does not need internet access to work so there shouldn't be worried that you will be locked out of your car if you are in a place with spotty service.

If you want to pair up your car with your iPhone, you must have the car manufacturers app so that you can set up your car key. During the first setup you out the iPhone in the cars NFC reader. iOS will recognize your car and redirect you to the car manufacturer's app to complete the process of setup. Automatic pairing with the car may fail, if the automatic pairing fails the manufacturer's app generates a PIN CODE which you can use to manually pair up your Car Key with the car.

Start your car

The digital key can be used to start your car without using the physical keys that come with it. Traditionally, we are used to using a physical key to start our cars. With the breakthrough in technology, we started using a fob (remote) to control some functions in our car. The digital key will revolutionalize how we interacted with our car.

The Car key can be used to start a car if the feature is supported by that car. To use this feature to start a car the digital key in your device must have been pair up with your car.

If your car supports the feature it will have an NFC reader or a wireless charging pad where you will place iPhone and it will authorize you to start your car.

Unlock your car

Digital hey has a whole host of functions it can be used to do after pairing it up with your car. Unlocking your car has become very easy than ever. After pairing up the Car key feature with your car your iPhone can be used to unlock your car. This feature uses NFC (Near Field Communication) to perform this function.

Unlocking your car with your iPhone, you simply tap your phone on the door handle. You can set it to require authentication so it will require Face ID or Touch ID after tapping your phone on the door handle or you can use it in the express mode which is the default mode after pairing up your car.

Set car key type

You can set the Car Key type when you share your digital key with someone. You may be wondering how you will share your car key if another person wants to use it. The digital key can be shared with another person. The digital car key can be shared via the messaging app.

The highlight of this function is that you can set the car key type you share with the person. Some access can be restricted when you share your car keys. You are able able to restrict some features such as acceleration limits, top speed limits, traction control, or stereo volume limits to ensure minimal distractions when driving. This feature can be particularly useful when you are lending your car to a teen driver. The digital key can be shared with up to five drivers.

The shared car key appears as a rich content in the message thread. To share your car key with someone

- Tap on the car key card in the wallet app.

- Then tap the - button in the upper right corner of the screen.
- Tap access to set whether to grant the driver unrestricted access to your vehicle or set the car key to Restricted Driving mode to limits some capabilities of the car.
- Tap the invite button and it takes you to a message thread.
- Type the name of the person or phone number then tap the send button.

Remove key access

As you are able to share your car key easily you are also able to revoke it easily. You may be worried that after sharing your digital key, will the driver continue to have access to your car. Don't woryy the shared key can be revoke if you no longer want the driver to have access to your car.

To revoke or remove the drivers car key access

- Open the wallet app on you iPhone
- Tap on the card for your car key

- Tap on more(-) button
- Tap on the name of the person you have the shared the digital key with.
- Tap revoke access for them to no longer have access to the car. This person receives a notification that their access has been revoked.

Power reserve

Battery life is often a concern on iPhone, you may be worried if you don't have access to electricity for long and your phone switched off. Battery drain does not necessarily mean that you the digital key. This is because power reserve, this makes your car key to work even after your phones battery is drained for regular use.

The power reserve mode lets your car keg to work for five hours after the phone may have been switched off.

CHAPTER THIRTEEN

Car play

Figure 13 Car play

Car play feature has around in iOS for quite some time now. In this new update, the car play feature is not left out when others to be lagging when other apps in the iOS 14 are receiving upgrades. Car play is Apple's feature which brings communication and entertainment functions to the built-in screen of your car using iPhone apps. The car play features are described by Apple as a smarter way, safer way to use the iPhone while in your car.

The platform is the best way to bring the iPhone experience to your car infotainment system. It is the ultimate co-pilot, it performs as many functions as your iPhone so that you can now use your iPhone apps safely. The car play feature let you perform functions which include getting turn by turn direction, make calls, send and receive messages, listen to music, and more. The platform also grants some certain apps for use in your car and makes good use of Siri letting you issue commands and listening to content without taking your eyes off the road.

The car play feature in iOS 14 receives an upgrade to further enhance the platform and added a whole host of other functionalities. In the new upgrade, you can add a wallpaper to the background, some new app categories, sharing your ETA with Siri, and a horizontal status bar for cars with a portrait view for a wider app view.

Though some car manufacturers have a somewhat smart built-in interface the car play feature lets you hands-on use your iPhone apps on the infotainment system that works with the car play feature. Apps that with the car play feature are installed on your iPhone, this all shows up when you connect your phone with the car on the car interface.

The car play feature does not replace the manufacturer's stock system and you can return to it with a tap anytime. To use the car play feature, your car must be compatible with it and there is some aftermarket stereo unit that works with the feature. The connection of your iPhone to your car can be made through a lightning USB cord or connected wirelessly.

Setting up your iPhone so that you can use the feature

Start your car and make sure Siri is on

Connect iPhone to your car by either using a lighting USB cord that will be plug into the USB port that is labeled with a car play or smartphone icon or if the car supports wireless car play, press and hold the voice command on your steering wheel. Make sure your stereo is in wireless or

Bluetooth mode. Then on your iPhone, go to settings > general > car play, tap available cars then select your car.

Wallpaper

In iOS 14 you are able to change the background of the car play interface. This means that you can change the background of the car play dashboard and will no longer have to look at the black wallpaper of the car play. Preloaded wallpapers are available that you can select from.

You cannot use customized wallpaper yet, but Apple has some small selection of colorful options to light up the screen of the interface that can change depending on the time of the day.

The appearance of the interface can also be set to either automatic or always dark. There is no option for always light for safety reasons when driving at night.

Changing the wallpaper of your car play

- Tap the setting app on the car play interface

- Tap wallpaper
- Tap on a wallpaper
- Tap set

  Changing the appearance of your interface

- Go to settings on the car play interface while your iPhone is connected and carplay is running
- Tap appearance
- Tap always dark or automatic. There is no option for always light maybe for safety reasons when driving at night.

New app categories

The car play has always supported some apps. These apps include apps, messaging apps, maps, radio, podcasts, audiobooks, and so on. These apps are installed on your iPhone and they will come up when the phone is connected to the car.

In iOS 14, some new categories of apps are added to further enhance the functions of the car play feature. The new categories of apps added to let you use apps such as EV Charging apps, Quick

Food Ordering apps, and Parking apps when the car play is running.

You can now use your favorite apps in these categories when the car play is running in your car.

Share ETA with Siri

The new iOS update comes with a feature where you can share your ETA (Estimated Arrival Time) with your personal contacts. When you are using Apple maps when driving you can share your journey status with your a person so that they when should expect you and where they will meet you. This is especially useful when you want to meet up with a person in the city and you hit traffic, sharing your ETA will make the person track your live location and gets an update about your arrival time.

You can share your ETA with your contact even if the person is not using iOS. If the recipient is not using iOS, they get the notification as a text message and get periodic alerts if your ETA changes substantially.

If the contact you share your ETA with is using iOS they get a rich notification which when opened they are redirected to Apple maps and able to track your live location.

To share your ETA on Apple maps

- Open Apple maps and set the destination you have to reach
- Tap on directions> Go
- Swipe up on the white tab and at the bottom of the screen to access the menu
- Tap on share ETA. Then select the contact you want to share ETA with (you can select more than one contact).

Horizontal status bar

The status bar of the car play shows vertically at the left-hand side of the interface. With this new update, users that have cars with a portrait display will now the option to move the Car play status bar to the bottom of the screen for a better layout and wider app view.

CHAPTER FOURTEEN

AirPods

Figure 14 AirPods

AirPods are known for their convenience, they make it easy for us when listening to our favorite podcasts, programs, or kinds of music. The earbuds also perform performs some whole host of other functions which include from double-tap access to Siri to pick up calls and a whole range of functionalities.

With iOS 14 AirPod and AirPod Pro even became better with better features thanks to some software-based enhancements Apple has added. When iOS is updated, the software-based enhancement will make the AirPod perform better. When the device you use the AirPod with is updated to iOS 14, the AirPods and AirPods Pro will gain some new features.

AirPod and Air Pod Pro have new features such as Spatial Audio, Headphone accommodation, Battery notifications, Automatic device switching. These software-based enhancements will further enhance the performance of AirPods and AirPods Pro.

Headphone accommodations

This is an accessibility feature for those who are hard a lot of hearing. When the device you connected your AirPods to is updated to ios 14, you will be able to access this feature. Apple makes AirPods be able to amplify soft sounds and adjust frequencies to make movies, calls, music, and sound more crisp and clear.

The headphone accommodations feature enhances audio to aid the people that are hard to hear, so they can be able to hear sound crisp and clear. In this new feature, there are various options available such as running audio for a balanced tone, Vocal range, or Brightness and you can also adjust soft sounds to become louder.

The headphone accommodation feature will help people with hearing issues. As part of the process of using the headphone accommodation accessibility, your iPhone will play different tones and sounds for you to determine what suits your hearing. This makes you customize the audio setup to meet your audio preferences.

AirPods Pro when the device you are using then with is updated to iOS 14, will an additional option to use the headphone accommodation credibility feature in the transparency mode and tweak the amount of ambient that passes through.

To access the headphone accommodation accessibility feature

Tap AirPods
Go to the Audio accessibility setting
Tap on the Headphone Accommodations to access the feature.

Spatial audio

This feature is for the AirPods Pro only. The spatial audio lets you be in the moment and makes the audio sound like it was coming right from your phone than from your AirPods. This feature provides a unique listening experience and brings a movie theater surround sound experience.

When watching movies or playing games that have surround sound on iPhones and you have updated your OS to iOS 14, you will be hearing your audio in surround sound if you are using AirPods. AirPods Pro plays back a virtual surround experience that is similar to those on Dolby Atmos for Headphones or DTS headphone:X software.

This spatial audio employs dynamic head tracking for the feature to work and bring you a movie like surround sound. The spatial audio is complicated in the mobile environment because the sound shifts as you shake your head or move the iPhone.

The accelerometer and gyroscope in both the AirPods Pro for dynamic head tracking. The data from the accelerometer is then combined with directional audio filters and changes in the frequency of audio to recreate an immersive audio experience like the surround sound in a movie theater. Since it is using dynamic head tracking, the surround sound channels are also adjusted as you move the head around.

This feature does not work on the regular AirPods but it only works on the AirPods Pro when your iPhone is running iOS 14.

Battery notifications

This is a new feature that comes with the update. iPhone and iPad will now be able to show when your AirPods and AirPods need charging. When listening to music to music and your AirPods and AirPods pro are getting low on battery and to be charged, your iPad and iPhone will now be able to let you know with a notification so they can be charged before it switches off completely.

Automatic device switching

With iOS 14 yourAirPods and AirPods Pro can now automatically switch devices seamlessly. Though AirPods and AirPods Pro features easy and quick switching where you are signed in with your iCloud account, this feature gets even better with iOS 14, in the new update your AirPods and AirPods Pro can now switch automatically between devices that are paired to the same iCloud account.

You can swap quickly between devices automatically if you are using the iOS 14 when listening to an audio on your iPhone and changes to watching a video on your Mac, AirPods will connect to the Mac automatically. When devices are switched you still have to change the audio output.

This feature works with the AirPods Pro and the second-generation AirPods. It is not compatible with the original AirPods.

AirPods Pro Motion API

Apple design a Motion API in iOS 14 which can be used by developers. The Motion can be used by developers to access user acceleration, orientation, and rotational rates.

Developers can take advantage of the motion API to build games that AirPods Pro owners can download. Access to the API (Application Programming Interface) can be a big deal for fitness apps, to track your movements or workouts.

Developers may develop a fitness app that automatically counts reps for activities where your head moves and your wrist barely does such as pull up or push up

And it can also be useful in gaming using data from your head motion. This will be a great addition to gaming too, developing games that use head motion such as a quiz app which will be able to tell if you nod yes or shake your head for no will be a good game.

CHAPTER FIFTEEN

Camera

Figure 15 Camera

With iOS 14 the camera app has come with better features to enhance how the iPhone camera works. IOS 14 has redefined the iPhone camera app. If most of your time on your iPhone is shooting videos and taking photos the new features for better filming and capturing experience. Even if you are just a casual photographer or videographer the new still benefits you with the changes to the default camera app on your iPhone.

The update has some new features but some are just some improvements on existing tools for improved functions. Also, some features which are once exclusive to newer iPhone models are brought to the older models when updated to iOS 14 so everyone can enjoy the same great features.

In iOS 14, a green and orange light indicators appear at the top of your screen, the green light indicates that the camera is in use and the orange indicator shows that your mic is in use. When you are taking your photos the green light indicator will appear and when you are recording the orange light indicator will appear. This feature can also as a privacy measure as you will know when some apps are actively using your camera and mics.

This means that if you see the orange or green indicators and you aren't aware of any needing your camera there may be an app that is secretly spying on you by accessing the camera. This change is part of Apple's push to give users more confidence about their privacy through transparency. If you find an app abusing your camera you can revoke their camera permission in the settings app.

To know which app uses your camera recently, you access the control center on your iPhone. So you can revoke their camera permission in the settings app if you don't want to uninstall or install them from your iPhone.

Improved shot to shot performance

Efficiency improvements and speed are added to the Camera app by Apple with iOS 14. Photos can now be captured up to 90 percent faster, at up to 4 frames per second than before. Apple says the time it takes to get the first shot after opening the app is 25 percent faster and capturing portraits is 15 percent faster shot to shot.

The improved performance is made possible by the updated software behind the camera so that it doesn't lag as much as it used to when taking a photo. There is also a feature where you can "prioritize faster shooting". Toggling on this feature in the Camera section of the Setting app will adapt image quality when you are rapidly pressing the shutter so you can make sure you don't miss any shot due to processing times.

Quick toggles in video mode

In iOS 14, all models of the iPhone can now quick toggles in the video mode if the device can run the iOS 14. The quick toggles feature comes with iOS 13.2, only the 2019 and later models of iPhone runs these feature. The quick toggle makes you be able to switch resolution and frame rate directly from the camera app.

All models of the iPhone can now use the quick toggle of the camera when updated to iOS 14. By simply tapping on the current video resolution or frame rate you can toggle between options.

This feature can also be used in the slo-mo mode to adjust resolution and frames per second If it does not work you need to go to settings, then camera, and toggle on "Video Format Control".

QuickTake videos on iPhone X models

The QuickTake videos feature introduced in iOS 13 for some models of which include iPhone 11, 11 Pro, 11 Pro Max, and SE (2nd generation). This QuickTake feature lets you take a quick video without switching to the video mode. The feature is intended to offer fast and easy to shoot a video while in the photo mode without switching back and forth between the two modes.

The feature has also to iPhone models when updated to iOS14. The QuickTake feature has been expanded to iPhone XS, XS Max, and XR too.

Mirrored photos captured on front camera

iOS 14 now has an option to take mirrored selfie images. When a selfie is taken, it will have been flipped on checking the Photos app but that is not always the best. Though photos can be flipped using the editing tools in Photos. A selfie image can be taken without going through the stress of flipping using the editing tools.

The front camera can now be set to capture mirrored images, so they look exactly like when you are capturing the image. To capture mirrored images

Go to settings and tap camera

Toggle on "Mirror Front Camera" under composition.

Exposure compensation control

In iOS 1r there is a new exposure compensation control. The new exposure compensation dial makes it easy to adjust the level of brightness of your photos making it lighter or darker. In the Camera app, when you opened the camera tool above the shutter button a new exposure compensation icon will appear. The icon is a circle with plus and minus, when the icon is tapped the dial can be swipe to the left and right to adjust exposure.

The dial is labeled with exposure value numbers, just like a mirrorless DLSR camera. A tiny opened appear at the top left part of the display when the dial is opened, the meter shows how much of a change you have made when you swipe the dial from right to left. When the tiny is tap the EV (exposure values) also shows.

CONCLUSION

Through the years Apple iOS has seen a ton of updates to improve its features and performances. This has seen Apple adding new features, functions, reshaping the look of the OS, and making everything more secure. The latest model of iOS sees some groundbreaking changes even if the changes the iOS looks completely different with the new type of widgets.

The new iOS 14 works with a broad range of iPhone models starting from the iPhone 6s that came out back in 2015 and the two generations of the iPhone SE. There is a guarantee of five years of updates from Apple using this software. The new iOS embraces widgets and reassess how you use your apps. The iOS 14 is refreshing and some new features we're added to the software for better performance.

Apple focuses on privacy makes the software incredibly user-friendly and mostly easy to explain putting back the focus on developers to consider what data they are collecting from us. You may not embrace all the changes to the iOS 14, they can coexist with the old way of using your iOS.

Take the iPhone experience everywhere with the new spatial audio on AirPods Pro that makes you hear surround sound and change your audio experience, the new translate app which can make to be in conversation with a person of another language is a great addition.

The software has redefined how you interact and use your phone, the iPhone can now be used as a car key to unlock and start your car. The OS also introduce tons of quality of life improvement which are some of the great added features to the iOS

The new features added by Apple I the iOS 14 is worth praising and the changes are worth it.

www.ingramcontent.com/pod-product-compliance
Lightning Source LLC
Chambersburg PA
CBHW070358220526
45467CB00001B/430